PLAN OF SEWERAGE, ETC.

REPORT

AND

PLAN OF SEWERAGE

FOR THE

CITY OF CHICAGO, ILLINOIS,

ADOPTED BY THE

BOARD OF SEWERAGE COMMISSIONERS

DECEMBER 31, 1855.

CHICAGO:
PRINTED AT THE OFFICE OF CHARLES SCOTT,
Corner of Clark and South Water Streets.

1855.

NOTICE.

SEWERAGE COMMISSIONERS' OFFICE,
Chicago, Ill., Dec. 31, 1855.

THE Board of Sewerage Commissioners hereby give notice that they have fixed upon the Plan of Sewerage specified in this publication, for the three Districts of the City, and that they will receive written objections to the same, at any time within thirty days of the date hereof.

W. B. OGDEN,
J. D. WEBSTER,
S. LIND,
Sewerage Commissioners.

(See section 10 of the Sewerage Law.)

REPORT AND PLAN.

TO THE HONORABLE MAYOR AND
COMMON COUNCIL OF THE CITY OF CHICAGO:

SEWERAGE COMMISSIONERS' OFFICE,
Chicago, Ill., Dec. 31, 1855.

THIS board, having adopted the plan of sewerage embodied in the accompanying report of their chief engineer, respectfully submit it to the judgment of your honorable body, and their fellow-citizens.

W. B. OGDEN,
J. D. WEBSTER,
S. LIND,
Sewerage Commissioners.

SEWERAGE COMMISSIONERS' OFFICE,
Chicago, Ill., Dec. 26, 1855.

GENTLEMEN:

The first question that presents itself in the investigation of any plan of sewerage for this city, is, *what is to be drained?*

The commissioners having already decided that the plan of sewerage to be devised should cover, at present, the territory included within Division street on the north, Reuben street on the west, North street on the south, and Lake Michigan on the east, it remains to decide what shall be admitted into the sewers.

As the main object of the sewers is to *improve and preserve the health of the city*, it is very obvious that all substances should be received into them which would have a contrary effect, if not drained off. As a consequence, all stagnant water, all liquids from kitchens and manufactories, and the contents of all privies, should be admitted into them.

With regard to *privies*, they are, as at present constructed and used in most parts of this city, abominations that should be swept away as speedily as possible. To construct the vaults as they should be, and maintain them in only a comparatively inoffensive condition, would be more expensive than to construct an entire system of sewerage for no other purpose, if the past experience of London and other large cities, is any guide for the future of Chicago. The general board of health, in their "minutes of information" relative to the "removal of soil water," etc., published in 1852, state, that after careful investigation, they are satisfied that the additional cost of removing the contents of privies, under the old system, together with the cost of constructing and maintaining cess-pools, beyond what it would be if cess-pools were abolished, and water-closets, draining directly into the common sewers adopted, is thirty shillings sterling per house per annum. If an estimate less than one-half of this, or three and one-half dollars per house per annum, should be applied to Chicago, then for 20,000 houses, or a population, say of 150,000, the saving would be $70,000 per annum, or the interest at seven per cent. on $1,000,000, more than enough to construct an efficient system of sewerage, for that number of inhabitants; provided, they were not spread over a greater proportional area than the population in the south district of the city now is, north of Harrison street.

What shall the sewage of the city be drained into? Four principal plans, besides modifications, have been proposed:

First. Into the river and branches directly, and thence into the lake.

Second. Directly into the lake.

Third. Into artificial reservoirs, to be thence pumped up and used as manure.

Fourth. Into the river, and thence by the proposed steamboat canal into the Illinois river.

The first plan has been adopted for the system now recommended. The reasons in its favor are, that it would allow the sewers to be constructed in such a manner as to take the utmost advantage of the natural facilities that the site of the city affords, and consequently, that the sewerage may be less in extent and cost.

The prominent objections are, that it would endanger the health of the city, especially during the warm, dry portions of the year, and that it might fill up the river so much as to obstruct its navigation. It is proposed *to remove the first*, by pouring into the river from the lake, a sufficient body of pure water to prevent offensive or injurious exhalations, by means which will hereafter be described. *The latter objection is believed to be groundless*, because the substances to be conveyed through the sewers into the river, could in no case be heavier than the soil of this vicinity, but would generally be much lighter. While these substances might, to some extent, be deposited there, when there is little or no current, they would, during seasons of rain and flood, be swept out by the same force that has hitherto preserved the depth of the river.

To many it is a matter of surprise that the river should maintain its depth, when there is so little current in it during most of the year; but those who have carefully studied this subject, see not only how the depth is preserved, but also how the spring, and other freshets, have been sufficient, from time to time, to make the upper parts of the south branch wider and deeper.

The objections to the second plan, or draining directly into the lake, are,

First. The greater length of sewers required, and consequently greater cost.

Second. The difficulty in stormy weather of preserving the outlets from injury, or from being obstructed from sand and ice.

Third. The supposed effect it might have on the water with which the citizens are supplied from the lake, if any of the outlets should be near the pumping engine.

The objections to the third plan, or draining the sewage into the reservoirs, and then pumping it up, to be used for agricultural purposes, are,

First. The great uncertainty about there being a demand for the sewage after it is pumped up, sufficient to pay for distributing it.

Second. The great evil that would necessarily result from a failure of the reservoirs through insufficiency of capacity, especially if the system of sewers leading to them should have their outlets too low to empty into the lake or river. If the reservoirs should be made so large as to place them beyond all doubt of having sufficient capacity, they would be very expensive, both on account of the labor and materials required in their construction, and the ground they would occupy.

Third. There would be danger to the health of the city during the prevalence of winds from the quarter in which the sewage might be used as a manure, especially, if only a few miles distant, and spread over a wide surface.

Should the time ever arrive *when the value of the sewage as a manure*, would be so great as to call for a system of drainage, permitting it to be saved for this purpose, the sewers now proposed would by no means be useless, but would be exceedingly valuable in providing for the surface water, and thus allowing the most economical construction possible, of a system of small sewers to drain simply what would be valuable as a manure, together with as much surface or other water as might from time to time be necessary for flushing purposes.

It is said by very respectable authority, (see the article " Drainage," recently published in the new edition of the Encyclopedia Britannia,) that *to drain a city perfectly*, a system of sewers should be constructed for the surface water, separate from that provided for other kinds of drainage.

Without, however, constructing a new system of sewers, the contents of those now proposed, might easily be collected by intercepting sewers, parallel, or nearly so, with the river and its branches, and then pumped into such other sewers or reservoirs as might be deemed most advisable.

With regard to the fourth plan, or draining into the proposed steamboat canal, which would divert a large and constantly flowing stream from Lake Michigan into the Illinois River, it is too remote a contingency to be relied upon for present purposes; besides, the cost of it, or any other similar channel in that direction, sufficient to drain off the sewage of the city, would be not only far more than the present sewerage law provides for, but more than would be necessary to construct sewers for five times the present population.

Should the proposed steamboat canal ever be made for commercial purposes, the plan now recommended would be about as well adapted to such a state of things, as it is to the present, making it necessary to abandon only the proposed method of supplying the South Branch with fresh water from the Lake, and to pump up from the new canal, or draw from the Des Plaines directly, flushing water for the West District, instead of obtaining it from the present canal at Bridgeport, as herein recommended.

How shall the receptacle for the sewage be reached? In most cities this can easily be done, by taking the shortest practicable line from each point to be drained, to the receptacle, through sewers that empty themselves by simple gravitation. In Chicago, however, the general level of the surface is so low, that a most careful and deliberate consideration is required in answering this question. During the last few years, much has been said and written, with regard to the *proper form* and *necessary declivity for sewers*. The experience of European cities has been sought with great interest by those who have recently been called upon to devise systems of drainage for the cities of this country.

It is not thought necessary or expedient here, to give an account of the conflicting views that have been expressed in London and elsewhere in England on this subject. The documents which contain them are voluminous. (Most of those published by Parliament may be found in this office, by those who wish to consult them.)

Some have contended that a declivity of one in two hundred was necessary to make ordinary sewers self-cleansing. This pre-supposes great irregularity in the depth, and consequently in the velocity of the current, which irregularity is generally found to exist in practice. In fact it is so great, that we know of no city or large town, with a system of sewers so perfect throughout as not to require, occasionally, either flushing or cleansing in some of its parts.

If, then, provision must be made for cleansing the sewers but occasionally, it becomes a matter of great importance, financially, as well as practically, to know how much should be expended, and to what extent natural advantages should be sacrificed in order to obtain a nearer approximation, merely, to a self-cleansing system, before encountering the risks and uncertainties of a general system of deep sewers in a wet soil. Besides the danger of *greatly increased cost of placing the sewers deep*, especially if the season should be rainy, the masonry in such sewers after being laid, would be, to say the least, as liable to get out of order as that in those laid nearer the surface, and in all probability much more so, because of the danger of the greater head of water finding its way through imperfect joints, either in the masonry or pipes, and ultimately of undermining the sewers. When once injured, deep sewers, in such a soil would be much more expensive and troublesome to repair, and a corresponding increase of inconvenience and annoyance would be felt by those who should depend upon the constant action of them for the drainage of their premises.

It is sometimes urged against deep sewers, requiring machinery for pumping, that the engines are liable to get out of repair, but as there should be duplicate machines kept constantly in order, this objection has not much force, in times of peace and health.

In time of war, however, when such machinery would be liable to be destroyed, or in a time of frightful pestilence, when men and means for making necessary repairs, might not be at command, the dependence of a large city upon engines for pumping up its sewage, might prove a terrible calamity. These are possible contingencies, and should therefore be considered.

It is known by actual experience, that a *five feet sewer, with a declivity of only one in twenty thousand,* filled half full with water, would have a current through it sufficiently rapid to move all substances that should be allowed to enter it from kitchens or water closets. Such a sewer filled to a uniform depth of six inches, with a declivity of one in twenty-five hundred, would have the same force of current. This is the greatest declivity it would be practicable to give to the main lines of sewers in Chicago, and at the same time have their outlets sufficiently high to discharge into the river, without the aid of mechanical power.

As in practice it might be impossible to *keep out of sewers improper substances,* provision must be made for cleansing them, either by hand work, flushing, or giving them ample declivities, and as already mentioned, the latter mode has not proved universally and perfectly effectual in any large town or city, so far as is now known. It would be impossible in this city to give to the sewers what is considered sufficient declivity to make them self-cleansing, without sinking them very low in some places, and pumping up their contents. In view of the practical objections to these conditions, which objections will be more fully considered toward the close of this report, it is now proposed to dispense with pumping up the sewage, and to give to the sewers such inclinations as the nature of the ground and the proposed modifications of the grades of the street will admit.

In order to *keep the sewers free from offensive deposits,* it is proposed to introduce a slight, but constant, current into the mains, from sources independent of the city water works, and to resort to flushing or cleansing by hand, whenever necessary, either in the branches or mains.

Both theory and practice show conclusively that the *circular form of sewers,* when not less than half filled, is the most efficient. When considerably less than half full, the egg-shaped sewer, modified in form so as to suit the depth of the current through it, is the most efficient. As in practice a uniform depth is never maintained, local circumstances must be carefully considered before a satisfactory decision with regard to the form can be arrived at.

In Chicago, two important circumstances bear with great weight on this subject. *The first* is, the necessity of keeping the top of the sewers as low as possible, to avoid expense in raising the grades of the streets; and *the second* is, the desirableness of giving to them the utmost efficiency in proportion to their cost and sectional area for discharging their contents during, and immediately after, heavy storms. For the latter purpose, the *circular form* is *undoubtedly the best*, while the former would require an oval with the flat side down. Where there is only a slight depth of current passing through the sewers, the egg shape, small end down, would be more efficient than the circular form for large ones of the same sectional area. Under all these circumstances, it is thought best here to reccommend the circular form as generally preferable, especially if bricks should be used in the construction of the sewers, and there is but little probability that, for most of them, any other material, procurable at the same cost, would answer so well.

In accordance with the foregoing views, *a system of sewerage for this city* has been devised and will now be described:

By referring to plan number one, it will be seen that the *sewers in the south district of the city* have their principal dividing or summit lines on State and Washington streets; that starting from these dividing lines, they discharge westwardly into the south branch, between North and Washington streets, northwardly from Washington street, into the main river, between Market street and the lake, and eastwardly into large mains on Michigan avenue, one of which empties into the river, and the other has its outlet into the lake, on Twelfth street; and, that small branch sewers run through the streets, which lie parallel with the summit lines, so that every lot may be reached.

In the north district it will be seen that three main lines, extending from Division street to the main river, and having their outlets on Rush, Clark and Franklin streets respectively, are proposed; also a main, having its outlet into the north branch on Chicago avenue.

All the intermediate streets between the mains, and those running east and west, it is proposed to drain by branches of different sizes, as shown on the plan, so that every lot may be reached the same as in the south district. It will be observed that no sewer has its outlet into the lake in the north district.

In the west district it will be seen that mains from Reuben street to the south and north branches, are proposed. For the present it is recommended that they be constructed only in Prairie, Randolph, Monroe and Van Buren streets, and in these, only so far as existing improvements may require them.

The streets and parts of streets intermediate between the mains, are to be drained by branches, as in the south and north districts.

It is recommended *that all the main sewers be circular*, five feet in diameter, except the one through Clark street, in the north district, which, on account of its probable future extension northwardly, it is proposed to make six feet in diameter.

In a few places, especially in Michigan avenue, in the south-west corner of the west district, and wherever else mains must evidently terminate, they might diminish towards their upper ends to four and three feet diameter.

The *largest branches*, it is proposed to make two feet, and the *smallest* fifteen and twelve inches in diameter. As already mentioned, the five feet sewers could have a declivity of two feet per mile. The two feet sewers not less than five feet, and the twelve and fifteen inch ones not less than ten feet per mile. The smaller sewers should enter the larger ones with curved junctions, somewhat elevated above the bottoms of the latter. The *mains to be built* eight inches thick, of brick, of the ordinary form, and in two rings or shells; the *larger branches* of wedge-formed bricks, four inches thick. The *smaller ones of earthenware glazed pipes*, if they can be obtained in sufficient quantities of good manufacture.

For the purpose of *supplying the mains with* a constant current of *fresh water*, and also the means of flushing both them and the branches, it is proposed to lay lines of flushing pipes, marked in red on plan number one. These are to be four feet in diameter, and laid in North, State and Washington streets, in the south district, Division street, in the north district, and Reuben street, in the west district.

In the south and north districts it is proposed to keep them filled, by pumping from the lake, to a height of eight or ten feet; and in the west district, by drawing directly from the canal at Bridgeport.

These *flushing pipes* are to be made *of oak plank* two inches thick, tongued and grooved, held together by iron hoops, and placed so low that the air would be always excluded from them, except when they might be intentionally emptied, either for inspection or repairs.

For the purpose of *keeping the water in the south branch fresh*, especially during the dry and warm season of the year, when there would be danger of sickness from putrid exhalations, it is proposed to construct a canal twenty feet wide and six feet deep at low water, between the lake and the south branch, through North street, and to protect its sides and bottom with oak timber and plank, and to cover its top with the same, so as not to interfere with the free use of the street.

By means of *an eighty horse engine and a wheel*, there could be driven through this canal four hundred cubic feet a second, or, in twenty four hours, thirty-four millions five hundred and sixty thousand (34,560,000) cubic feet, a quantity equal to all the water there now is in the river, between the lake and North street.

With such a *supply of fresh water from the lake*, it would be impossible for the south branch to become dangerous to the health of the city, even with all the sewerage of its present and prospective population.

With regard to *the outlets of the sewers*, it is proposed to place them so that the bottom of the interior surface of the mains, would be six inches below the present surface of the lake, or eighteen inches above low water level of 1847; and to place the bottom of the two feet sewers, six inches higher, or about the level of the present surface of the lake.

As the lake, unaffected by winds, was, in 1838, twenty-four inches higher than the proposed outlets of the two feet sewers, these would then be full, and the five feet mains half full. This, if permanent, would be objectionable, as it would make it troublesome to cleanse out the smaller sewers especially; but, as the past general average level of the lake, according to the best information we can get, has been below that of the present season, about two-thirds of the time, it is believed that the benefit of having the outlets as low as now proposed, would, at mean and low water, be much more than an equivalent for the inconvenience of such an arrangement at high water.

It is proposed *to receive the surface water* not only from lots, but streets *into the sewers*, so long as the latter shall be capable of receiving it.

To prevent sand, or other heavy substances, on the surface of the streets, *from washing into the sewers*, and thus clogging them, there is to be, at the corner of each block, *a catch basin*, similar to those used in New York, Boston and Philadelphia. These must be emptied of their solid contents at intervals of time, depending very much upon the locality in which they are placed; in New York, some have to be emptied every three weeks, and some not once in three years. These are so constructed as to prevent offensive gasses from escaping into the streets.

For the purpose of having *access to the sewers, either to flush, cleanse, examine or repair them,* it is proposed to have entrances covered with manhole plates, at all the intersections of streets, and at intermediate points, where the blocks are more than four hundred feet long. These entrance places would serve as convenient points at which to place stop-boards or valves for flushing purposes.

The *smaller sewers should be provided,* at intervals of about one hundred feet between the entrance places, *with movable covers,* about three feet long, so that in case of stoppage the sewer could be opened and cleansed without being broken.

In order that *every branch sewer* may be supplied *with flushing water* whenever necessary, without depending upon the city water works, it is proposed to connect all the branch sewers together, at their summits, wherever the streets run in continuous lines, when, by means of stops or valves, it will be practicable to turn a stream of water in any direction.

In some portions of the city included within the territory it is now proposed to drain, *more streets will have to be laid out,* before it can be properly improved, and then only can the details of sewerage be judiciously arranged for such portions.

It is a question of great practical importance *whether the branch sewers should be laid through the alleys or the streets.* If alleys were systematically laid out through every block, for most purposes it would be better to lay the branch sewers through them ; but where one block has an alley through it, and the next none, it would be necessary to lay a sewer in the street between. In such a case, to lay an additional sewer through the alley would be to the public a useless expense, although to individuals it might be very convenient.

When *the ordinary privy,* with its *cess-pool in the back yard, is abolished,* and water-closets in the house are substituted, and the bottom of the cellar made high enough, a sewer in the main street would be preferable to one through the alley.

It would not be possible, in the low streets, with the system now recommended, *to have deep cellars,* or any not liable to occasional flooding by backwater from the sewers, *unless guarded by self-acting valves.* The latter, however, is a difficulty from which no city, built on a comparatively flat site, is wholly exempt ; and no system of sewerage providing at all for the surface water, could, without enormous expense, do away with it entirely. If, however, individuals should wish to have cellars as low as the sewers would drain them at ordinary times, they could protect them against the influx of back water of heavy storms, or when flushing might be resorted to, to cleanse the main or branch sewers, by the use of a simple valve, similar in construction to the one used in common pumps, and so efficient in the eastern cities, where cellars are often made with their bottoms several feet below high tide.

As all arrangements of this kind necessarily depend, for their efficiency, upon being faithfully kept in order, and upon all improper substances being excluded from the house drains, the city should in no wise be responsible for them.

The *main sewers would not be large enough to receive the surface water in extraordinary storms*, that is, when over an inch an hour should fall in summer, or a much less quantity when the ground is covered with melting snow. To obviate the difficulty that would be felt at such a time, and to prevent, as far as possible, collections of water from remaining in the streets, it is proposed to give to the latter, grades, inclining constantly, though irregularly, towards the river or its branches, or the lake; and to depend upon slight undulations in the gutters, to turn the surface water into the catch-basins at the corners of the blocks.

In order to make proper *provision for receiving the surface water*, it will be necessary ultimately to lay large main sewers through each street running east and west, in the west district, if the city should grow as rapidly as is expected. In that case the small branches now proposed would have to be taken up; but even if the material in them should be wholly destroyed, the saving of the interest on the difference of first cost, would, in five years, be more than sufficient to pay for them; but, it is believed that the materials in them, though liable to injury by being removed, need not be wholly lost. The value of five or more years of experience would be very important in determining, from the working of those first laid, what the sizes of the other mains should be; for it is impossible to tell, without actual experience, how much of surface water, in time of heavy storms, would find its way immediately into the sewers, and how much would be absorbed by the soil.

In order *to give to the sewer a sufficient depth of earth over them*, which should be nowhere less than two feet, it will be necessary to raise the grades of the streets above their present level an average of eighteen inches, for twenty-five hundred feet, westward from the river branches, in the west district, a great part of which has already been provided for by the grades the city council has recently established.

In the *extreme south-west* part of the territory embraced in the plan, the *ground is too low*, by nearly four feet, *to cover the proposed sewers* to a sufficient depth; but as the drainage of that part of the city might be taken southward into the river, instead of eastward, as shown on the plan, and a large portion of the filling saved; and as this part of the city may not be improved for several years to come, it is deemed sufficient for present purposes to state the general depth of filling that would be required by the system now proposed.

This low district has nearly a triangular shape, and extends as far north as Tyler street, and as far east as Halsted street, and the depth of filling required to raise the streets over it, would average about two feet.

So far as any *new grades have been adopted* for the *south district,* they are sufficiently high for the proposed sewers. South of Madison street, and north of the line of Tyler street, where no grades have been established, the streets should be raised an average of eighteen inches nearly for two thousand feet from the south branch. South of the line of Tyler street this filling would extend but one thousand feet from the river, and be of the same average depth.

In the *north district,* between Division street and Chicago avenue, and extending from the proposed north branch canal to an average distance eastward of thirty-three hundred feet, there would be an *average fill of eighteen inches.* South of Chicago avenue and west of Dearborn street, the average depth of filling would be two feet. On Franklin street, where it is proposed to lay a main, it would average three feet.

If the streets should be raised much above the natural surface of the ground, it would be *necessary to protect the yards* and rear lots from being flooded by back water from the street sewers *in very heavy storms,* by a valve arrangement similar to the one necessary for cellars, provided a temporary flood would at such a time be considered a serious injury.

The *main sewers nearest* and parallel with *the lake* in the south and north districts, could *easily be relieved* of excessive storm water, *by waste wiers* placed about three feet above their bottoms, and discharging into lateral conduits connected by the shortest possible lines with the lake.

For the *mains and larger sewers, the best material,* as already mentioned, is believed to be brick. The two feet *branch sewers* should have *glazed bottoms* to the height of three inches. This mode of building sewers has been practiced in England, and no doubt could be carried out here. For the *small ones,* wherever fifteen inches or less in diameter may be used, *earthenware glazed pipes* would be preferable, if they could be obtained at a cost not over fifty per cent. higher than they are made for in England.

A successful *manufacture of these pipes* would be of *immense importance to this city*, as great numbers of them must be used, especially for house drainage; and there does not appear to be any good reason why such a manufacture should not be established.

In consequence of the impossibility of determining at this time the cost of different kinds of materials, the *estimates* will be *based upon the ruling prices* of the *past season* for bricks, and upon what the makers of pipes in other places offer to furnish them at. It is proposed to build into the main and branch brick sewers, *junctions for every house, and* every *lot* likely to be occupied by a house, or to need draining, to avoid the necessity of breaking the masonry whenever permission to enter the sewers may be asked.

The foregoing description should be considered as giving but the general features of the plan, and not the details.

Further information may show the propriety of modifying either the location or sizes of some of the mains, branches, and flushing pipes.

The following *estimates of cost* do not cover the sewerage for all the territory embraced in the plan, but merely so much as is considered necessary for present purposes, together with the general arrangements required for changing the water in the river, and for flushing the sewers as far as they may be laid down: That is, the *territory covered by the estimate* is bounded by Chicago avenue, LaSalle, Erie and Franklin streets, the main river and north branch, Carroll, Prairie, Sangamon, Madison, Halsted and Harrison streets. (See blue line on plan.)

South District.

4,950	Lineal feet of large sized sewer, at..	$4 25	$21,047 50
34,940	" " 2nd " "	1 25	43,662 50
38,630	" " 3rd " "	1 00	38,630 00
12,330	" " flushing (4 ft.) pipes....	1 10	13,563 00
336	Catch basins and fixtures.........	50	16,800 00
30	Gates on line of flushing pipes....	20	600 00
120	Entrances, with man-hole plates, and fixtures for flushing at each point...	30	3,600 00
	Proportion of engine and conduit for driving water into the south branch,		20,000 00
			$157,893 00

North District.

9,020	Lineal feet of large sized sewer, at..	$5 25	$47,355 00
21,280	" " 2nd " " ..	1 25	26,600 00
25,790	" " 3rd " " ..	1 00	25,790 00
7,920	" " " for flushing pipes	1 00	7,920 00
2,870	" " flushing (4 ft.) pipes	1 10	3,157 00
356	Catch basins and fixtures........	50	17,800 00
10	Gates on line of flushing pipes....	20	200 00
90	Entrances with man-hole plates, and fixtures for flushing at each point..	30	2,700 00
	Proportion of engine and conduit for driving water into south branch and main river....................		20,000 00
	Engine house, and fixtures for pumping flushing water...............		5,000 00
			$156,522 00

West District.

9,600	Lineal feet of 5 ft. sewer, at......	$5 00	$48,000 00
21,560	" " 2nd size "	1 25	26,950 00
34,125	" " 3rd " "	1 00	34,125 00
22,960	" " " for flushing.......	1 00	22,960 00
15,860	" " flushing pipes........	1 10	17,446 00
325	Catch basins and connections......	50	16,250 00
20	Gates on line of flushing pipes....	20	400 00
90	Entrances, with man-hole plates, and fixtures for flushing at each point..	30	2,700 00
	Proportion of engine and conduit for driving water into the south branch,		20,000 00
			$188,831 00

If the foregoing estimates of the cost of providing sewers for that portion of the city which now needs it, should be taken as a criterion for the whole territory embraced within Division, Reuben and North streets and the lake, then the *entire cost of sewering* the latter would be about $2,300,000.

It will be perceived, that no allowance has been made for driving fresh water from the lake into the north branch.

This has been omitted because it is proposed to drain into it, at present, but a very small portion of the west district. Ultimately, however, such a provision would, no doubt, be found necessary.

For the purpose of *purifying* as much of the *north branch* as possible, it is believed that the necessary canal should be located as far north as Center street.

No plan especially adapted to *subsoil drainage* has been devised, although the subject has been considered.

It is believed that the provision herein recommended, for drawing off immediately all water that may fall on the surface, except in time of extraordinary storms, would, in a great measure, do away with the necessity of subsoil drainage so far as healthfulness is concerned.

The funds provided for, by the sewerage law, are not sufficient to meet what should be done at once for house drainage. It would therefore be useless to delay this report, to prepare a plan of complete subsoil drainage, which if adopted, could not be carried out without further legislative authority.

The *drainage of cellars has been considered.* When their bottoms are not too low, they could be drained into the proposed sewers, provided the precautions against being flooded by excessive storm-water already mentioned, should be used.

This, in most localities, would allow of a cellar six feet deep, if the first floor was placed two feet above the sidewalk. Deeper cellars could not be provided for, without pumping up the contents of the sewers, or raising the grades of the streets higher.

Were a system of sewers designed with especial reference to the drainage of such cellars as are now desired, it is by no means certain, that it would be sufficient to meet the demands of a few years hence.

The *sewers of New York and Boston* have generally been laid with reference to what was supposed to be, when they were made, ample cellar drainage, but already there are to be found in both cities "*two story cellars,*" which cannot be drained into the common sewers.

The main object of the *sewerage law*, is evidently *to provide for the healthfulness of the city;* deep cellars, though of great value already, in some parts of it, especially on Water, Lake and Randolph streets, and very convenient in others, are not essential to health, but probably diminish rather than increase it. The fact, however, that the drainage of deep cellars may not be provided for at the public expense, need not prevent the owners of them from joining together and draining them, at an annual cost of less than ten dollars for each twenty feet front. A practical proof that this estimate is sufficient, is, that well-constructed cellars in this city, are already drained *by hand*, at this small cost.

Perhaps this may be the most proper place to say a few words with reference to a system of drainage strongly recommended by some, because of its success on a small scale, in certain localities. The system is that of *submerged sewers emptying into the lake or river.* The supposed advantage being, that the substances admitted into the sewers, would be kept constantly in a fluid state, would not emit such offensive odors, and would be easily driven out by a head of water. It is very evident, however, that the mere fact of a sewer being submerged, would not increase its capacity for discharging, and that if it were long, with but a slight head available for flushing, it would be impossible to create a current sufficiently strong to cleanse it. In that case it must sooner or later fill up. Such sewers have been laid again and again in London and Paris, and other cities, and have received universal condemnation, except for short lines with an effective available head. When such sewers become filled up, they are often abandoned, in consequence not only of the expense and danger to life from cleansing them, but of the intolerable nuisance such an operation would create.

The different communications, thirty-nine in number, that have been received by the Commissioners, *proposing plans for draining* this city, have been examined. Some of them are very able and interesting papers, and though not followed in the system now recommended, they have been very valuable in pointing out matters of great local importance.

In conclusion, it will be stated, that while very able and competent engineers differ essentially in their views as to the best mode of draining such cities as Chicago, it would not be safe to rely merely upon numerical estimates of costs of construction or maintenance solely, in deciding such a question.

After due consideration of all estimates that can be based upon reliable data, important points still remain to be decided by judgment, and cannot be reached by mere computation. As the opinions of well-informed men have ever differed on important subjects, it is not to be expected that they will be unanimous on this.

After weighing carefully all the arguments for and against different plans of draining any locality, the only course is to decide upon that which promises to combine as many advantages, with as few disadvantages, as possible.

With this view of the subject, it is believed that *the plan herein recommended should be adopted*, and for the following reasons :

First. The first *cost of construction*, would be as little as by any other mode equally efficient.

Second. The *cost of maintenance* would be as little as by any other equally efficient plan. The apparent complication of the system of flushing, is really no greater than should be provided for in other plans, though not required for use so often. The necessity of a regular and periodic attention to the matter, would insure greater exemption from offensive gasses, than a less frequent and more irregular system, or something nearer what is termed a self-acting one. As to a really self-acting system, requiring no attention or supervision, it is nowhere to be found in artificial drainage of cities, any more than in other branches of art, necessary for the ordinary wants or comforts of life.

Third. The *healthfulness of the city* would at once be *greatly benefitted* by removing from the surface all stagnant water, if, at the same time, a vigilant police should prevent animal and vegetable substances from being exposed to decay in the streets or lots.

Fourth. The *liability to serious interruption would be less*, than on any other plan ; and if repairs should be needed, they could be more easily made. The importance of having every part of so extensive a system easy of access, in case repairs should be needed, can hardly be over-estimated.

Fifth. *Emptying the sewage through the river would carry it further out into the lake*, and consequently place it more beyond the reach of being offensive than any other plan, except that of transporting it to the country, to be used as manure.

Sixth. *Clean water* constantly poured into the river and main sewers, in sufficient quantities, would be *the best de-odorizer* and absorbent of noxious gases it would be practicable to obtain.

Seventh. *No uncertain sources of supply*, or irregular modes of action *would be depended upon for purifying the river or flushing the sewers ;* and no useless expense would be incurred, by requiring the city water works to pump more than ten times the necessary height, admitting that they were capable of supplying such quantities as would be requisite. This would be impossible, if, at the same time the demands upon them for domestic and manufacturing purposes should be met.

Eighth. *The system is capable of being extended in every direction*, as the growth of the city may demand it ; while each sewer built, would be capable of "beneficial use, independent of its further extension," as required by section eight of the Sewerage Law.

Ninth. *The deficient inclinations of sewers is compensated for by introducing a constant stream into the mains*, and by means of *flushing the branches;* especially if the latter should be wholly or partly glazed.

The means of introducing a stream of fresh water into not only every main, but almost every branch throughout its length, are exceedingly available in Chicago. The low and flat site of the city, so unfavorable with regard to giving to the sewers the most desirable inclination, is of the greatest advantage in affording a remedy for the evils of such deficiency.

For the most reliable information concerning the fluctuations in the surface of the lake, we are indebted to Messrs. GOODING and GUTHRIE, of the Illinois and Michigan Canal, Col. J. D. GRAHAM, U. S. Army, and I. A. LAPHAM, Esq., of Milwaukee.

Messrs. WM. H. CLARKE and JOHN REID, Civil Engineers, and WM. GAMBLE, Clerk of the Board, have rendered valuable and satisfactory assistance, in the different labors necessary for the preparation of this report and plan.

Respectfully submitted,

E. S. CHESBROUGH,

Chief Engineer of the Board of Sewerage Commissioners.

TO THE SEWERAGE COMMISSIONERS,

Of the City of Chicago, Illinois.

www.ingramcontent.com/pod-product-compliance
Lightning Source LLC
LaVergne TN
LVHW011146110826
845150LV00008B/2547

* 9 7 8 1 4 1 8 1 9 1 7 2 6 *